YOUR
FIRST YEAR
IN DATA
SCIENCE

DENNIS SALGUERO
@DSWITHDENNIS

First Edition

Some names, places, and details have been modified to protect my clients and projects.

Artificial intelligence was not used in the creation of this book.

Beta Readers:
Yan Guo
Jia Guo

God has never promised me a fast plan or even a good plan. But there is a plan.

TABLE OF CONTENTS

Preface   9

Acknowledgments   11

Introduction   13

Why Data Science?   17

Transition Into Data Science   23
    What Is Data Science?   23
    Transition From A Student   28
    Transition From Another Career 30
    Impostor Syndrome   33

Six Pillars of Your First Year   35
    Methodology   35
    Data Sourcing & Preparation   35
    Model Search   36
    Model Deployment   36
    Stakeholder Management   36
    Data Science Team Dynamics   36
    Your Team Leaders   37
    The Data Scientist Mindset   38

Methodology   41
    A Word On Visualizations   46

Data Sourcing & Preparation   47
    Data Sourcing   48
    Repeatable & Consistent   50
    Data Enrichment   52
    Data Labeling   55
    A Word On Creativity   57

Model Search   59
    Champion Model Search   59
    Iterative Development Process 62
    A Word On Bias   63
    A Word On Notebooks   66

Model Deployment   67
    Prediction Runtime   68

Application Development Layers   71
Graphical User Interfaces   73
Putting It All Together   75
Model Portability   77
Model Maintenance   77
The Cloud   78
Other Languages   79
Stakeholder Management   81
Communication   81
Education   83
Ideation   84
Data Security & Discretion   86
Human Validation   87
Demonstrating Impact   89
Data Science Team Dynamics   93
Team Meetings   94
Project Management   95
Remote vs. In-Office   97
Pair Programming   98
Model Review   100
Get Out There   103

Preface

This is not my first time writing a book.

In the late 90s and early 2000s, I had just graduated college and went directly into self-employment, riding the "dot-com wave", and getting as many consulting contracts as I could. In between contracts, I would sustain myself by also doing technical editing and writing programming books.

Eventually, I became a co-author for a book on Access 2000 programming - you can still find it on Amazon - and it taught me a lot about how a book comes together. But mostly, it also taught about just how *hard* it is to write a book.

After the writing, the editing, the multiple technical reviews, and everything else that a manuscript has to go through, I was exhausted! I promised myself that I would never write a book ever again!

Yet, here I am.

If you know me in person or see my content on social media, then you know how passionate I am about data science. I have decided to make this field my life's work, and my mastery, and I'm always looking for ways to extend my skills in this field. I have been fortunate to have a long career and I find myself doing more teaching and mentoring at this stage in life.

This book represents the next natural step in that evolution and that is what motivated me to write it. I look forward to sharing this knowledge with you and, hopefully, bringing more talent to the field that has brought me so much joy. It hasn't been easy to put this book together, but this is simply one of those things I have done with passion and that makes it a *little* easier.

Also, I have decided to write this book in the first person. While this may be slightly unconventional, I think it's the best way for me to express some of my thoughts and speak from a position of mentor and teacher, which I think you will enjoy.

If this book piques your interest, I invite you to join me on social media, where I share even more data science content. I also host live streams on a regular basis and those are really fun to put together. You can find me on all the major platforms @dswithdennis

Acknowledgments

First, I want to thank my family for all their love & support. Mom, Dad, Karla, George, Martin & Elizabeth, and the extended Pouliot family. I also can't forget my dog, Keno, for always keeping me company when I have my double espresso and he has his pup cup at Starbucks.

I also want to thank my friends throughout the years: Ekka, Janet, Kevin, Al, and Bobby.

I want to thank all of my colleagues who have shaped my career in some way. Marc, Josh, Ali, Ed, Phil, Ben, Justin, Thomas, Edward, Jia, Yanhong, Yan, Pedro, Fernando, and all the other people I've had the honor to work with.

Finally, I also want to thank all of my followers on my social media platforms. When I first started my accounts on May 1, 2024, I had never really been a big user of social media. This was my first time aggressively trying to grow an account. I figured that about 500 followers after my first few months was going to be a success.

As I write this in early July 2024 - only about 9 weeks since starting my accounts from zero - I have more than 10,000 followers. Unbelievable! It's been amazing to see how many people consume my content and I would have never written this book without your inspiration.

Introduction

I am a very blessed man. I am convinced that one of life's greatest victories is to be able to find work that you enjoy and maintain the skill & time to dedicate yourself to that work.

Find a job you enjoy and you'll never work a day in your life.

I know that is a trite statement, but I do believe in it. I haven't worked in years! I enjoy data science and it's all I've been thinking about for the greater part of the last 12 years.

This career has given me many amazing things. I've been able to work with some very talented colleagues, I've traveled the world, I've owned a Rolex Daytona and a Porsche 911 and I now have the lifestyle where I own my time and pick projects that I enjoy.

Another benefit that I have received is that I've been able to form a data science community through my social media platforms and I get to meet and chat with data science enthusiasts (of all levels) from all around the world. That brings me an immense amount of joy and it's something I hope to continue doing until the day I die.

Through these conversations with my followers, I can gather common themes that affect the greater data science community. Some of these you can probably guess, like searching for jobs & internships or what degrees to study. I get those questions daily - sometimes multiple times per day - and I have created some online resources to help students with those queries.

Another theme that became clear to me was that there is a wide gap between how you do data science in an academic setting versus how it's done in the real world.

You landed the data science job. Now what?

Let's look at the standard course of study for most data-oriented curriculums around the world:

- The course gives you the model to use. A curriculum usually includes things like regression methods or neural networks. So the model is, effectively, dictated to you. No need to conduct your own experiments to find an appropriate model.

- The assignment gives you the project idea. No need to come up with your own project or be creative, it's already selected for you.

- The professor gives you the data to use in a nice, clean & compact format. No need to source your data or worry about data quality.

But the real world is nothing like that! In fact, it's quite the opposite of all those three factors. Most first-year data scientists don't know that this transition is required, much less how to navigate through it.

I hope to fill that gap with this book.

I have been able to guide other first-year data scientists through this process but this is the first time I have put those steps down on paper. I hope this helps you, the reader, with that process.

Or perhaps you are also a data science leader and want to know how to improve your process when onboarding a new data scientist. You might find some guidance in this book too.

Either way, I hope this book helps set you up for a long, successful career in the world of data science.

CHAPTER ONE

Why Data Science?

Let me be clear: data science is hard. And, just like anything else in life, it's even harder when you don't enjoy it.

Throughout this book, you will often hear me speak about how I believe that data science is a learned profession. I equate it to being a doctor or a lawyer where you - being new to the field - rely on the existing professionals to show you how things are done in the real world. A doctor can't possibly learn every single health procedure in school alone. A lawyer has to go through additional mentoring to be a litigation lawyer who stands in front of a judge in a courtroom.

And in both cases, no one becomes a doctor or a lawyer by mistake, they bring a certain passion to it. Would you want your surgeon to be someone who cares about & is fascinated by the human body? Or someone who became a surgeon because of the money?

There are many articles, postings, infographics, etc. that tout data science as being a great job for remote workers or as being "the sexiest job of the 21st century" or ways to "hack" the data scientist interview and get a job. I feel that these postings degrade the profession and downplay the work that goes into becoming a data scientist.

That's why, when someone first approaches me about learning data science and advancing in the field, I usually find a sneaky way to ask, "Why?". What exactly are the motivations that are pushing you into data science? How do you know that data science is for you?

I did not start as a data scientist straight out of school. I don't even have a technical degree; my undergraduate studies are in International Business from The George Washington University, a degree of study that I am proud of.

I don't come from money. My parents sacrificed a lot for my sister & I while growing up. I remember my parents bought us an IBM PS/2 Model 30 when I was still in elementary school and it made me handy with computers. I would sit there and program in BASIC, mostly copying code out of the back of Family Computing magazine that I had checked out from the local library so that I could play some games on my machine.

I kept up with my programming skills as I made my way through primary school & high school. I didn't take many formal classes, I just felt like it was an interesting hobby that I would use at some point in the future.

I had to work my way through college. That is easier said than done when you're studying in Washington, D.C., the land of the free . . . interns! The majority of internships don't pay any money in D.C. but I couldn't afford to go that route, I needed a paycheck!

I was extremely fortunate to land a job near campus, being the tech support guy for a dial-up ISP provider for this new thing called "the internet". They gave me flexible hours and weekend work so that I could still keep a full-time class schedule while giving me just enough money to keep my studies going.

I was able to take that experience - along with my programming skills - and start working in the IT/IS world straight out of college. More importantly, I graduated right around the time when dynamic, data-driven websites were starting to become popular. I started by working with languages like Cold Fusion and VBScript, which is now commonly referred to as "classic" ASP. A year or so later, I was a (very) early adopter of Microsoft's .NET framework and started working in VB.Net, C#, and ASP.Net.

All of this work provided a very good living for me. I was always able to find work and was employed at some very interesting companies. I did a lot of work in the entertainment field so I was able to attend some cool sporting events and concerts. As a consultant, I was able to travel the world and help some really interesting clients.

It was a good life . . . but I wasn't happy. At core, I was still just doing the work for the money. I remember that I had a good friend at the time, Barry, who also worked in the same programming languages that I did. And it's only natural to compare yourself to your friends. Barry enjoyed going to programming conferences, pursuing certifications in the field, and advancing his skills as a developer. I didn't have a lot of interest in that, I wasn't passionate about it to that same degree. Inside, I was unhappy. I was happier with the paycheck than I was with the work and I didn't want to continue like that.

Right around this time, I was introduced to the book, *Mastery* by Robert Greene. If you are not familiar with it, it's a book that advocates finding the work that you want to dedicate yourself to, while also realizing that mastery takes time and commitment to your craft. But if you follow your passion and allow yourself to be consumed by your mastery, then nothing but good will come of it.

I use audiobooks quite a bit and I put this book on my phone while I would walk back and forth from the gym. The book had a massive impact on me and made me do some personal reflection on what exactly I wanted to dedicate my life to.

Through that reflection, I realized that I've always been pretty good with numbers, particularly statistics & probabilities. And I still enjoyed some aspects of writing code. Hey, that sounds a lot like what a data scientist does!

And that is how I chose my current path. True to the book, I feel the happiest I ever have at any point in my career because I am passionate about the field and, quite frankly, a lot of the work in data science comes naturally to me now that I have mastered most aspects of data science.

But despite the mastery, I am still in awe of what we are able to do at the intersection of math, code, and probabilities. Every time that I execute some code or figure out a neural network, there is just an overwhelming feeling of magic that is not always easy to explain. In many ways, I don't want to think about it too much, I just want to keep on discovering new things within data science. As I mentioned previously, this isn't work for me, I'm just having fun!

I'd like to think that this passion comes across in this book and my video content. I think it's a part of why my subscriber base has grown quickly. I hope that this enthusiasm brings other people into the field for the same reasons!

I share all of this to say that I hope you find your passion. In one of my videos, I ask all my data science students to read *Mastery* and maybe you will find equal inspiration in that book.

I hope you end up pursuing something that will bring you happiness from your life's work. Maybe that's data science, maybe it's not. Either way, it's OK! But you should have a pretty good reason for why you're doing data science, otherwise, you won't be happy over the long term and that's no way to live.

I will close this section with the following: If you talk to an Olympic athlete, say a javelin athlete, they can show you the mechanics of how they throw the javelin and instruct you. But you, as a beginner, might have a million questions that the athlete isn't always going to address.

You may wonder why the wrist and hand placement has to be a certain way. Why is it at the front of the javelin and not the rear? How do you know how many steps to run? Your coach may not always explain those things or even be clear on why. The athlete just replies, "I don't know, I haven't thought about it in years, it's just the way I've always done it".

I have done my best to provide examples and illustrations for all of the concepts that I discuss here. But I acknowledge that I may have missed some nuances that I do just based on my years of experience. You may very well have questions where I would have to tell you, "I don't know, I haven't thought about it in years, it's just the way I've always done it"

CHAPTER TWO

Transition Into Data Science

There are many ways to transition into a career in data science and no one way is better than the others. While this book focuses on two ways - coming out of school and coming in from another career - I think these thoughts still apply to anyone coming into the field. If you bring both your passion and your best efforts, you should be able to find success in the field and maintain a long career in data science.

What Is Data Science?

So that we can start on equal footing, I want to establish some of the basic beliefs that I hold about data science and those that my experience has shown me have held up over time. Whenever you are in doubt about something that I present in this book, or perhaps at some point later in your career, you should be able to trace the answer back to one of the tenets that I present in this chapter.

If you are not consistently running experiments with your data then it's not data science

Working in this field, I have always thought of myself as a scientist above all else. I'm not a programmer, I'm not a statistician, I'm not a math genius. I am a scientist.

I encourage you to do the same because this mindset will always center you when you're unsure of what to do next, both in projects and in your career. My experience has shown me that data science is the continuous application of the scientific method to create algorithms to solve problems that are presented to you. When in doubt, always refer back to your application of the scientific method.

I will present my interpretation of the scientific method so that we can work from a common understanding. But please recognize that there are variations on all these steps and you are welcome to seek out other interpretations.

1. **Observations:** What data can you collect and what does it tell you?

2. **Hypothesis:** Do the observations cause different questions and doubts? Great, those are a good start for a hypothesis

3. **Experiments:** What are some of the experiments that we can do on our data to support the hypothesis?

4. **Interpret Results:** What do the experiments conclude about the hypothesis? Do the experiments create results that lead to more questions? Can the results be used to formulate a future hypothesis?

> *Pure data science is the removal of bias from your data and creating a process for data-driven decisions*

As I have matured in my data science career, I have become more & more hyper-aware of the bias that exists all around us. I intend that statement both for my own bias and those external to my thoughts.

There are many forms of bias - too many for me to list here - but awareness is half the battle. But bias can be useful. Where bias helps is in the realization that we should always be questioning ourselves and our approach to data. What am I missing here? What are the gaps in my data? Are there any other data points that I should be considering here? All of these are healthy questions to ask throughout your career.

The way I usually encapsulate this concept with my clients is "Let the data speak to you, not the other way around". Are you truly allowing the data to inform you & drive decisions? Or are you allowing some form of bias to shape your view? When you allow bias into your process then you are speaking for the data and no longer making data-driven decisions.

This won't be easy for you during your first year in data science. I concede that maybe you won't even understand this topic as you are reading through it now. It's one of those items that you will have to develop a "feel" for and I'm positive that this will make more sense to you as you grow into your data science career. You will also find mentions of bias throughout this book because I feel that it's an important topic.

> *Data science is a learned profession - like a doctor or lawyer - that requires on-the-job training, even after the academic portion of your career is completed.*

I know that I have already stated this tenet earlier. But since this is, essentially, the whole premise for this book, it bears repeating here.

This means that you should be completely comfortable admitting your knowledge gaps as you grow in your career. You simply will not have all the answers and no one expects you to have them either! Realizing that you *don't* know something is a sign of wisdom and you should stop and take the time to learn what you need before proceeding further. That's completely OK!

When you are searching for your first few roles in data science, you should be paying close attention to the people who will be above you in your reporting line. Not only should you look at them as your supervisor, but these people will also be your mentors that will affect your career and how quickly you learn & progress as a data science professional.

To be clear, no one is going to drag you along the learning curve for data science. In many cases, it will be up to you - and you alone - to speak up when you want to learn and need additional help in your journey. The resources and the people will be there, but you will need to reach out and appropriately utilize them. And that process starts when you are interviewing for a new data science role.

You should be asking questions to see how well you get along with your superiors and how effective they are as teachers. If the situation doesn't feel right to you, then don't accept the offer. That's a perfectly acceptable choice to make.

In turn, I hope that later in your career you will also take the time to teach and mentor people new to the data science field. It is up to us - as a profession - to help each other and establish a solid foundation for the growth and betterment of the world of data science as a whole.

I am certainly in the "seasoned" stage of my career. And while I still have one more CEO, CTO or Director of AI position left in me, this feels like an appropriate time to write down my thoughts and bring more people into the field. Hence, my content creation on social media and efforts like this book. Not only do I enjoy this process, but I also view it as my duty to help data science grow and I hope you will do the same when the time is right for you.

> *"The first guy through the wall, he always gets bloody"*
> *- Moneyball*

If you are truly interested in the field of data science, then you have most likely already seen the movie *Moneyball*. If you have not, I highly recommend you watch it; the movie is essentially about applying data science to American baseball.

Towards the end of the movie, the Brad Pitt character is meeting with the owner of the Boston Red Sox. Pitt laments about how hard it was to convince people that his plan would work and how everyone doubted him. The owner admires what has happened with Pitt's team and tells him, "The first guy through the wall, he always gets bloody".

I've loved that quote from the moment I heard it and I have used it in many, many meetings throughout my career, both with my data science staff and stakeholders. This concept may not hold weight with you right now. But as you grow more & more in your career, meet with more stakeholders, and deliver more projects, this saying is going to come to life for you.

There will be moments when you will need to have difficult conversations because you are going to be presenting thoughts & ideas that break the norms that the people around you are accustomed to.

People are creatures of habits (read: bias) and those can be difficult to break. You will face resistance, you will get people that tell you that you're wrong, and you will get people that say, "We can't do that" or "We don't work that way".

But you will have to stand firm because - assuming you've done your job correctly - if it's what the data says to do then it's the right thing to do. You will have to develop skills to help you effectively present your conclusions and eventually "sell" those solutions through the final implementation.

These conversations and the ensuing wins will not come easy. You should wear these victories as a badge of honor. Are you bringing *real* value to your organization as a data scientist if you just prove what they already know?

Data science is about bringing innovation and new approaches to your business. And, yes, that means you're going to be breaking down some walls and getting a little bloody in the process. Chin up and eyes forward - you're going to be just fine.

Transition From A Student

If you're moving straight from school and into the workforce for your first full-time position, you're going to be on multiple learning curves.

You're going to have to learn how to collaborate as part of a bigger team. Not just within data science, but with related fields such as IT teams, documentation teams, your executive leadership, etc.

You're going to have to learn how to navigate your team's existing codebase and projects that are in flight. It's rare to have a brand new project just waiting for you on your start date, you'll most likely join something that's already in progress.

You're going to have to learn how to navigate additional items that come with a new job, such as health insurance, retirement plans, and human resources policies & requirements. All of those "adult" things are a critical part of the process.

Finally, you might also be joining a company and/or industry that you know very little about. You will have to do the work, on your own time, to learn about the history, financials, and current events for your industry and integrate yourself into the business side of your new company.

It's a lot to take in!

But if you were smart enough to get the job, then I believe that you are smart enough to navigate this process. Stay calm, take a deep breath, you can do this! Here are some items that I have shared with my new colleagues that I believe are useful:

- **Be Patient:** Your biggest source of pressure during this time will be your own mind. Give yourself the grace to take your time in this process. Realize that it's not easy, that you have a lot of things to learn, but that you're also making progress. You are *not* learning and working nearly as slowly as you think you are!

- **Carry a Notebook:** I have always found it useful to carry a notebook with me any time I am with a new client. Not only will it help you remember things more clearly, but it also helps to add a little more professional "polish" to your approach. It demonstrates your interest in the work and in doing a good job all around. Taking notes always helps me feel more prepared when I'm ready to sit down and do some work on my own for my new team.

- **"Slow is fast":** Whenever I am feeling stressed on a project, I always repeat this mantra to myself and I think it will help you a lot in your first year. In most situations, you are usually much better off slowing down, taking your time, and turning in a quality deliverable. Take the time to slow down your brain, think about your next task, and move in a deliberate manner. That is much faster than rushing, making a mistake, and then having to redo an item that wasn't very good. Take the extra five minutes now so it doesn't cost you an extra hour later.

- **Speak Up:** In my career, I have found that your colleagues are usually more than happy to answer questions about their work. A good professional can appreciate when someone is trying to ask good questions to educate themselves. Plus it demonstrates that you are coachable and able to take direction.

  Coming straight from school, you might not have that confidence level quite yet, but I assure you that your new colleagues want to hear from you. Don't be afraid to speak up, ask questions, get clarification. No one expects you to have all the answers right away; asking questions might help them feel better too!

Transition From Another Career

It is difficult to write this section generically because people come to data science from all sorts of backgrounds and experience levels, so I really can't target a single profile to speak to. I will just generalize that you are someone with at least a couple of years of work experience who decided to make a move to data science.

The good news is that the generalization works because there are so many facets to the data science development cycle, that you will likely find a place for you to adapt your skills into a good data science team.

You should also note that a good data science executive will welcome diverse backgrounds into their teams. The varied backgrounds also bring diversity of thought which is invaluable during ideation with stakeholders and collaboration in the data science team room. Don't be shy about sharing your background and previous work experience during the hiring process and once you're on a team. It probably helps in more ways than you realize!

Here are some common, professional backgrounds I have come across and where I have seen them have success in a data science transition:

- **Software Development:** As a data science project matures, it converges on what looks exactly like a software development project. A well-executed algorithm should exist in a stand-alone library that is deployed as a greater part of an enterprise application effort. That library should look & act just like any other library that someone with software development experience has worked with in the past.
  Or maybe you're a software developer who has a lot of experience creating API layers that require security and session management. That is a valuable skill to bring to a data science team that needs help with that "last mile" effort between their finished product and the end users.

- **Consulting:** Due to its growing nature, data science remains very dynamic in terms of the problems that a data science team is tasked to work on. Most data science teams work at the enterprise level; one day you're working with internal financials, the next day sales reports, the day after that on human resources data, and so on. If you have a consulting background, you will bring the project flexibility that this approach requires. You're already accustomed to jumping from project to project. Now you'll be doing it across departments and datasets.

Also, most seasoned consultants usually have some formal stakeholder management training and that can be a valuable service to a data science team, particularly during the initial stages when you are trying to define a use case to solve.

- **Business/Data Analyst:** I have known many data analysts over the years and the best ones can easily find a home on a data science team. The strength you bring is in the ability to interpret results and make algorithms mean something to stakeholders. This is a step that is commonly overlooked on data science teams and may very well be the reason they bring you on board. If you can display a comfort in producing both visualizations and text that can explain the results of a project in business language (not math or code) then you will make a smooth transition into a data science team.

- **Statistics/Math:** Having a background in statistics and/or math is valuable for a data science team. However, you can help yourself the most by looking at this from a meta-level. Most data science projects rely on existing libraries to handle most of the math and formulas that are needed. I don't think you'll be asked to write many (any?) formulas from scratch.

  But you will find parallels between the concepts and theories that you already know and put you back into your comfort zone. For example, your new data science colleagues won't say "Type I" or "Type II" errors, they'll call them "false positives" and "false negatives". It won't be a z-test, it will be "A/B Testing". You get the idea.

I also want to make sure that I am not discounting the intangibles that already being in the workforce has brought you. Most people underestimate the number of conversations and meetings with the business side that data science requires. That's not always an easy transition for someone new to both data science and working full-time. The communication and maturity that your "time in the trenches" has given you will only add to the ease of your transition into your new data science team.

Impostor Syndrome

I want to close out this chapter with a brief discussion on impostor syndrome. I don't claim to have any background in psychology or mental health. But I have been fortunate to mentor many young people in multiple corporate environments and I can't begin to tell you how often this topic comes up.

It comes up often enough that it felt appropriate to include in this book. I would support anyone who wants to seek professional mental health assistance, that's a positive thing. In place of that, I can only share my anecdotal experience here.

From my conversations with colleagues who approached me with help on this, I think that the feelings of impostor syndrome come down to two things: preparation and fear.

Data science can be an intimidating career choice. The processes and answers are not always clear so it's hard to feel like you are on the right path at any given time. The other reality is that *nearly everyone* in this field feels that way, you are far from alone in this! Your colleagues feel just like you do. I have been doing this work for well over a decade and I'm *still* humbled by the fact that data science is only making educated guesses!

That is why I try to break down the data science process as much as possible in this book so that you can see that there are still some basic tasks that you can focus on as part of your integration into the field. When in doubt, always come back to the scientific method and do whatever the next step dictates. You can rarely go wrong with that approach to data science.

So, take a nice, long, deep breath and congratulate yourself! By taking the steps to read this book - and maybe even do some of the exercises I present here - you have already made yourself more prepared than you were prior to opening this text. I firmly believe that the more time you take to practice & prepare - with your own data science studies that you own from end to end - the better prepared you will feel on your first day on the job.

By actually doing the work of a data scientist, even by yourself with no one reading your work, you have greatly reduced the probability of impostor syndrome creeping into your thoughts.

Finally, I have had many life experiences over the years - some harder than others - that have taught me a lot about myself and about fear. Again, I'm not an expert, but I can't help but think that fear plays a role in impostor syndrome.

The funny thing that I have found about fear is that it's not nearly as dramatic or fatal as we make it out to be in our minds. I think that some aspects of it can actually be motivating. The mantra that I repeat to myself is, "I don't have to be *fearless*, I just have to be a little *less* afraid than the other person". All of us are afraid in some way and that includes the colleagues that are part of your new team. A feeling of fear just shows that you are human, not weak or some other negative sentiment.

I believe that it's healthy to have some fear in your life; you just need to keep it to a level a little less than you think you can handle and it will serve you well in the future.

CHAPTER THREE

Six Pillars of Your First Year

You are going to learn a lot of things during your first year as a data scientist. Honestly, you're always going to be learning something while working in this field, even beyond your first year, that's part of the fun!

In this chapter, I present the six pillars that will, most likely, make up most of your learning for the first year. They are not presented in any order, none are more important than the other. These are just part of the tools that I think will come to the forefront during your transition to the industry. I provide a brief intro in this chapter and then you can read the respective chapters to do a deep dive into each one.

Methodology

Data science is not an algorithm or a result. It's the *process* to get to that algorithm or result. I will present you with a methodology to get started.

Data Sourcing & Preparation

Real-world data is not clean and it will most likely come from different sources. That's quite different from academia! Be prepared by doing the coding example I present in this chapter.

Model Search

You will have to create and implement your own process for how to find the appropriate model for your particular project. In this chapter, I present an approach to help you with that.

Model Deployment

You will no longer be measured by academic grades. You will be measured by your ability to complete algorithms and deploy them while also providing some sort of measurable impact. In this section, I'll help you get started with both of those items.

Stakeholder Management

The real world of data science requires a lot more communication than most people realize. While that is impossible to present in a single chapter (or even a single book!) I attempt to provide you with enough information to get started with your new colleagues.

Data Science Team Dynamics

You will have to learn how to work within a greater data science team. That's certainly not difficult, in fact, it's quite fun. Here I share some of the things that have helped me and how I run my own data science teams.

I hope that reading through these six pillars will give you an accurate representation of what your first year will hold for you. I have found them useful in training my new data scientists and I sincerely hope they do the same for you.

There are two additional items that I have to include in this chapter because I think they are also critical to your career growth.

Your Team Leaders

If we agree that data science is a learned profession, then who is going to teach you? Well, that is going to be a function of the team leaders that you will be reporting to in your first year of data science.

Depending on the size of the organization you join, there will be varying levels of skill on the team. There might be a Level 1 data scientist, a lead data scientist, a senior data scientist, and then a chief data scientist or some combination of all these levels. These same team members will play a role in shaping your data science career now and in the future.

What does that mean for you? It means that you should be taking your time during the interview process and learn as much as you can about each team. How big is the team? How long have they been with the company? What do their LinkedIn profiles look like? How many projects are they initiating versus how many are they maintaining? What will be the initial tasks they give you? Are there other data science teams within the greater organization?

Also, recall the previous discussion on how your team might have to break some biases and other beliefs along the way. How effective has your team been at this? Do you have strong leadership - particularly the chief data scientist - to push your projects forward with the rest of the company? What is the reputation that your team has among the business stakeholders?

If it sounds like I'm recommending that you need to do some interviewing of the team yourself, that's not too far off the mark. You should have a pretty good idea of the answers to all of these questions because, in many ways, it defines what you will be learning in the role. A doctor wouldn't join a hospital with a bad reputation, a lawyer would never join a particular firm without some research first. You should apply the same critical eye to the team you will be working for.

The Data Scientist Mindset

I can't close out a chapter like this without including a bit about the mindset that you will develop as a data scientist. Note that I didn't say you *should* develop, I said you *will* develop. It is only natural that the thought patterns you use in your data science work will start to creep into your daily thinking.

You will find yourself using more of a critical eye on the articles you read, the claims companies make, your expenses, etc. You will also find more creative ways to think about the world in numbers and ranges rather than how things appear in the natural. Just like a doctor who exercises every morning to take care of their body or a lawyer who soundly manages their personal affairs, you will find yourself being a data scientist in everything you do.

To illustrate, the second half of 2024 has been a season of building up my social media presence. That's not something I've done before, I haven't been a big user of social media. There's a lot of information out there, lots of people that want to sell you services and tools.

But instead of investing in that, I applied my own data science mindset, experimenting with different forms of content, and different days & times that I post. I even ran focus groups with some of my friends & old colleagues to critique my audio & video settings; I produced more than 10 testing videos before I ever posted one live item to any social media account. I pestered my friends & family for feedback. People told me that hashtags would not work, so I tested them, and the data I collected showed they *did* work. I continue to use this approach as I grow my accounts and help grow my audience.

You will start to see numbers all around you. You will start to think about how real-world items can be represented numerically for a model. You will start to use generative artificial intelligence tools for more tasks. The world will become a playground for your modeling. It can really give you a whole new perspective on everything around you!

Your mindset - in everything you do - will become research, predict, test, and then examine the test results and act accordingly. Hey, that sounds a lot like the scientific method!

I cannot present you with a discussion on the data scientist mindset without also mentioning bias. I'm sure you've learned about the different forms of bias in your data science courses. And I also have some further discussion about bias in other parts of this book.

But for this particular section, I'm referring to Researcher Bias or said another way, *your* bias.

> *I am open to debate on this.*
> *I am happy to be wrong on that.*
> *But, of course, I could be wrong.*

A positive sign of a sound scientific mind & approach is the realization that we are all just making very educated, but still variable, decisions based on the result of our work and experiments. Some of that work may be shaded - unintentionally, or otherwise - by our own bias. Some of that bias may be hard to detect because they can be errors of omission ("Oh, that will never work!") that can silently creep into our thought patterns.

I use the three phrases above when working with my teams because I want to always keep my mind open to new ideas, and points-of-view different than my own and I want to encourage the people around me to speak up if they see things differently.

I encourage you to do the same to reduce the probability of introducing your own bias into projects. Talk ideas out with other people, be unafraid to declare that you are wrong and pick a new direction for your experiments, keep your methodology open for review. These are all positive choices that will reflect in your work and increase collaboration on your team.

As you can see from all of the items presented in this chapter, not all data science involves you sitting down to write code. As long as you keep developing your skills and working more & more through the scientific method, you will find yourself maintaining more of this mindset in your daily life. It's at that point that you can start thinking of yourself as a complete data scientist because your professional skills have crossed over into your life skills as well.

CHAPTER FOUR

Methodology

If I can take all of my first-year data scientists and have them only learn one thing from me, then methodology would be it. It is the foundation upon which all of the data science projects I have ever executed are built and it brings many benefits.

> *Algorithms don't give you the right answer. They only produce the most probable answer!*

First, we have to start by accepting that algorithms, as advanced as they can feel, are not all-knowing, sentient beings. Far from it. They are just math formulas that we (as humans) can utilize in unique ways.

As a data scientist, you may very well already know this. But I'm presenting it here because you will be repeating that same sentence to your stakeholders in the future, guaranteed.

Since we accept that the answers are only the most probable, then we also have to accept that algorithms will (not *might*) make mistakes and be wrong at some point in the future. Even your best-performing algorithms will give you a bad result from time to time. Your stakeholders will then look to you for answers and you will have to decide if changes need to be made.

Everything that I have just presented to you can be effectively managed - both in the present and in the future - with a solid methodology process. The way you create and deploy your algorithms needs to follow a consistent plan that sets you up for the highest probability of success.

In addition, the mistakes from your algorithm may also present a risk to your company, both as a financial risk and a reputation risk. There may even be a legal risk where you will be required to demonstrate, and maybe even defend, your methodology under any number of circumstances.

If you can follow and document a methodology and provide the best code you can, then you have fulfilled your role as a data scientist for your organization.

Equally important, having a consistent and documented methodology adds to your professional approach to data science. Having a process that can stand up to peer review - by your team and other parties - is an integral part of the scientific method and a process I encourage you to embrace.

You are welcome to adopt your own methodology, there are many frameworks out there. Here, I am presenting what I use and it has served me well on many projects.

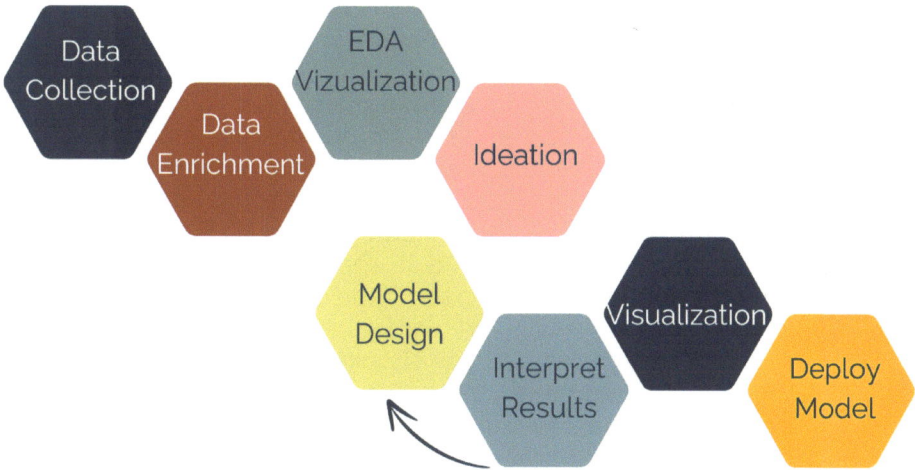

I believe that most parts of this framework are self-explanatory. But the following are some general notes that I think are key to understanding this chart.

- Any data science lifecycle usually presents multiple opportunities for visualizations and you'll note that I advocate for at least two of them. These stages are fantastic for educating your data science team, illustrating results to stakeholders, and for further **Ideation** when your work indicates that the data is demonstrating something contrary to current beliefs (bias).

  It's a healthy practice to stop looking at your project as a never-ending series of columns & rows and take the time to visualize different portions and either reinforce or reset your current plans. Other than some impact on your overall timeline, you really can't go wrong by having multiple visualization stages in your methodology.

- There is a loop indicated between your **Model Design** and **Interpret Results** phases - this is where the bulk of the work lies! This is to allow for the experimentation and tuning that your model will go through and you should invest as much time as you need to get through these two phases.

Realize that a lot is happening here and you'll have to define the sub-steps involved in these phases. For example, you'll notice that I do not have an explicit mention of human validation in this methodology. I would include that as part of your results interpretation and it should be part of the feedback that carries back into Model Design. But, of course, that will vary with your individual project & team practices and you should adjust accordingly. I also include more discussion on this topic later in this book.

- Having said that, make sure you don't rush to get to Model Design and take your time in the first four phases. The **Ideation** phase should provide you with a clear business case that your algorithmic efforts are going to address and the right data that you need to get started. Do not proceed with both significant stakeholder involvement - and their approval - and a clear direction of where your project is headed.

One of the better books that I have read recently is *Thinking In Bets* by Annie Duke. While she covers a lot of interesting items in the book, my biggest takeaway from her work is the separation of your decision-making process from the result. Just because you had a bad *result*, doesn't mean that your *process* was bad. In turn, a good result does not always indicate a good process - you may have just been a recipient of good luck!

This is another thought pattern that I use to emphasize the use of methodology in data science because we should all be trying to eliminate both the positive and negative aspects of variance (luck) in our work. The methodology represents our own process and we need to separate that from the results we will experience in the future from the algorithm.

What does this look like in practice? Consider the following . . .

> *"Don't try to impress me with your algorithm results. Show me that you followed a good methodology and I will defend your results to anyone who asks."*

The above is a statement that I share with everyone who works on one of my teams. I am willing to separate the results from the process by making sure that we have a statistically sound and defensible methodology that creates our results. If I, as the data science leader, sign off on your process, then I will always speak up for our team in stakeholder meetings and I'm also willing to accept when the algorithm makes mistakes once implemented.

At the risk of sounding like I have my own "Conjoined Triangles of Success", the following is a chart that I also share with all data scientists that work for me.

	Good Model Result	Bad Model Result
Good Methodology	**"Sleep Well At Night"** **Desired State**	**"It Happens"** **Negative Variance**
Bad/No Methodology	**"Luckbox"** **Positive Variance**	**"Unemployment Line"** **Irresponsible**

45

I don't know about you, but I like being able to sleep undisturbed at night. Maintaining a strong methodology in your projects helps greatly with that! That is the desired state that we should all be aiming for.

By the same token, we should all be looking to avoid the unemployment line because there should never be an excuse for creating a model that has poor methodology behind it. The poor results are the most likely and that's no longer being a scientist, you're just plain irresponsible at that stage.

A Word On Visualizations

Entire books have been written on visualizations and I recommend that you pick one up to further educate yourself. Clearly, I am a fan of them during the data science development process.

However, I do need to emphasize that you should recognize a line between the visualizations that you use for algorithm development & discussion within your data science team versus the visualizations that you use for your stakeholders. Those visualizations are *not* the same!

Resist the temptation to show your stakeholders cluster diagrams, formulas, distributions, confusion matrices, correlation diagrams . . . well, you get the idea. The visualizations for those meetings should be centered around the business and the resulting impact that your algorithms will create. Those diagrams should look nothing like the ones I mentioned previously.

You most likely won't be alone in this. There will be team members to help you. But it behooves you to adhere to the practice early since you don't want to lose your stakeholder to math formulas. Alas, no one stays for those meetings!

CHAPTER FIVE

Data Sourcing & Preparation

If I had to pick one data science task that is the most different between academia and the real world, this one may very well be it! You will find a large chasm between how you did data preparation in your studies versus how you will do it as a professional. You will also see that this portion of any data science project tends to be the most time-consuming.

In a lot of academic settings, maybe you never even had to source & prepare your data. The data was likely handed to you in a nice clean & compact format. In most of your projects, you most likely went straight from receiving your data and right into model creation for the given assignment. And, if we're being honest with each other, data science education has practically institutionalized the datasets we've all learned on. How many data scientists do you know who *haven't* done work on the iris dataset? Or Titanic? Or MNIST?

But don't despair. With a little bit of work and your coding skills, you will be able to make the adjustments to start working with datasets within your organization. I think that developing these skills is an excellent way to ease your transition into writing professional code and I routinely give these tasks to my first-year data scientists.

You'll note this chapter contains more technical terms than the others and also has some coding exercises. I love writing in a conversational style throughout this book, so the slightly more technical tone in this chapter is clearly intentional.

Your primary focus should be to create processes that are repeatable & consistent while also thinking about ways to enrich your data with an eye toward modeling. I hope the technical exercises in this chapter will assist you with that goal.

Data Sourcing

The first adjustment that you will have to make is the sourcing of your data. You will no longer be able to have data neatly handed to you, you will have to go out and engage in some hand-to-hand combat and collect your data!

This stage is also where you will have most of your conversations with your stakeholders. They should know the answers to most of the questions that you will ask of them for the data locations, file types, and things like that. View it as an opportunity to build a relationship with your stakeholders. This is only the first time of many when you will need their help.

However, I have always found it beneficial to take a "trust but verify" approach to what you are told in these conversations. Ask for samples of the data, visit the actual file locations, do test runs on importing unstructured data, and so on. Doing this type of initial exploration - maybe even *before* you decide to accept the project - can help you identify gaps in the data. You might find problematic items that maybe even the stakeholders didn't know about. That's all part of why we take the steps to verify.

You can break down your overall approach into two simple steps. You need to know where to *pull* the data and you need to know where to *store* the data. This is a problem where it's convenient to work backward, so start with the storage side.

Where are you going to put the data? Where can your team best access it for further processing? How many records do you need for modeling? How much memory storage are you going to need as the data stands now? What are your projections for the growth of that storage over time? Do you have enough storage for both scenarios? You should have answers to all these questions before moving forward to obtaining the data.

Now that you have a place to put your data, where are you going to get it? A big challenge is that you will need to be prepared to pull from a variety of sources and through a variety of methods. Some data is going to come from a database with ODBC, other data is going to be a JSON file via email, or it could be data you have to pull from a PDF file, stored in a filesystem, that you read via an API.

These are just some examples. But, no matter what your particular setup is, you will still have to answer: What is the advantage of each connection method? Will you have enough network throughput for that? How long will the code run for?

If you have not started your first job as a data scientist, then this is a good time to start developing the coding skills necessary to be able to connect to all of these data sources.

Also, keep in mind that you will have to work closely with your stakeholders to obtain a data dictionary for each of these sources. After all, we've only obtained the data at this stage, we still need to *understand* the data. A well-prepared data dictionary will help with this task.

Repeatable & Consistent

I have yet to come across a data science class project that requires you to update and maintain your dataset. Most classes are simply not long enough for them to show you how to work with dynamic datasets. Well, in the real world, you're almost always working with dynamic data. You will have to not only obtain your data (as described in the previous section) but you will also have to do it in a way that is repeatable and consistent. That means that you will have to:

- **Create** new records in your data as they become available.

- **Read** records via queries and collect the data you need for modeling.

- **Update** the records that you already have, as needed.

- **Delete** the records that need to be excluded from your production data.

  But you should know that deletes aren't actually carried out, you will most likely be doing this via a logical delete with a flag in your database.

For those of you with a "classical" programming background, you can probably already see that I did a poor job of hiding the classic CRUD operations (create, read, update, delete) that you may already know. But that's also because that is what is required for most of your data operations in data science.

This means that you should approach your data tasks with a programming mindset that accounts for all of these items. It's not difficult, it's just unlikely that you were exposed to a lot of this in your data science studies.

Here's an exercise for you to walk through; you can choose to simply think about the components for this or you are welcome to write out the code for this, it will be fun:

1. Make sure that you carry out all of the following tasks in a dedicated class that is accessed via a main entry point, not a script.

2. Source a dataset for weather information; there are multiple free APIs online for this.

3. Write code to access the API and store the next 10 days of weather predictions. Just get some simple predictions like temperature and rainfall. You will also want to include metadata like the city, province, country, and time of year.

   Store this in any format of your choice. But if you want to be specific about it - and prepare for working with APIs - go ahead and store it in JSON format.

4. After you have some data, go back to previous days that you stored *predictions* for and get the *actual* weather information.

5. Let's assume that we only care about the previous 45 days' worth of weather data, so use a logical delete to exclude anything longer than that.

6. In a completely separate class, write a query that pulls the last X days of weather data. X can be any value between 1 and 45.

   You want this as an independent class to simulate the business logic layer that you will use. You usually do not want to mix queries like this (and their respective methods) in the same classes that handle your data collection.

7. Use your favorite mechanism to have all of the above run on a schedule that collects the same data every single day at the same time of day.

When you are finished you will have two classes. One for your data collection (the API connection) and another for your business logic layer that returns data to your users (or application). And you will also have an automated schedule to run this.

I think that if you take the time to write out this task, you will be in a position to handle a lot of the data requests that you will face in the real world. The data will be different of course, but the pattern will be the same and prepare you for your first day.

By the way, if all of this sounds suspiciously close to data pipeline/engineering tasks, that's because it is. I have seen many data scientists kick off a successful career by starting as a data pipeline specialist on a team and working their way up. It gets you working on a team while also writing code in the process. Those are two very valuable skills to add to your resume and will serve you well in the future.

Data Enrichment

You're going to find data enrichment expressed in any number of ways. It can be referred to as "data wrangling", "data munging", "data transforms", "feature engineering" or other terms out there. Try to not get too caught up in those terms and stay focused on how you can help your model best.

> *Never accept data at face value. There are always additional features you can extract in a statistically defensible manner.*

Thinking back on my own data science studies, the datasets that I learned from were pretty simple. Not only was the data straightforward, but it was also fairly "compact" because it needed to fit quickly and easily on the school file server or some other convenient way to move it around.

But the funny thing about data is that the more you work with it, the more you start to see it as having the two dimensions of depth *and* width. You may already have a sense of depth; in simple terms, it's the number of records that are available to you. Maybe you even did some work with deep learning approaches in your studies.

It's that second portion, the width, that is addressed by the work you do in the data enrichment phase. In short, this is where you start to develop additional fields that you can add to your data to aid your modeling process. Those additional fields can add whole a new level of dimensionality when you start to model with your data.

The following are some illustrated examples for you to think about.

- **Timestamps:** At first glance, there aren't too many things special about a timestamp. In most tables, you can use them as audit trails or for simple things like that.

  But what if we were to think a little bit more about them? What if you start to break them down further? Consider a timestamp that you find in a table that contains retail sales transactions.

  - From the time you can also get whether it was an AM or PM transaction.

  - From the hour, you can bucket a sale as being a Morning, Afternoon, or Evening sale.

- From the date you can get the day of the week that the sale was made

- From the day of the week, you can also get whether it was a weekday or weekend sale.

- From the date you can get seasonal information like Winter, Spring, Summer, or Fall

- And the list goes on . . . and this is only 1 field in our initial data!

- **Weather:** Recall the previous exercise on creating a pipeline for weather data. While weather data *alone* may not be important, it does make for a good way to enrich your data. What if you end up working for a company that does a lot of consumer sales, like an event venue or maybe a car dealership? Sure, you could just focus on the internal data, I'm sure you will have plenty available. But wouldn't you also be able to increase the width of that data if you brought in weather information? Or what about macroeconomic information? There are datasets out there that allow you to take your information to a completely new level when you incorporate it into your models.

- **Creative approaches:** Data enrichment also allows you to bring a new perspective to a project. Models you work on may be influenced by other factors that are *not* in your internal data or need another view.

A few years ago, I read a whitepaper on stock market predictions and managing your portfolio with automated trades. That in itself is not unique, there are plenty of whitepapers on that same topic. But this whitepaper didn't take an approach to look at fancy indicators or sentiment in the news - that stuff is pretty standard. Instead, this paper looked at the Bid/Ask tables and the depth of the orders to make predictions - a very unique approach! By exercising some sound creativity on your part, you may find a novel approach to a problem you are trying to solve.

Remember that these are just some examples to get you started. A timestamp field is not the only field that can be extended. The weather is only one of many sets you can use for enrichment. And yes, the timestamp examples can be considered feature engineering, the weather example is data wrangling, etc. But don't get caught up in all that. The important part is to recognize the value of this phase and how it increases the width of your data.

There is no single exercise that will cover all of your data enrichment needs. It is a skill that will grow as you can put more real-world experience on your resume. In my view, this is one of the fun aspects of data science. It's one of those phases where you can just sit and *think* through a problem.

Data Labeling

I have some bad news for you: the world is not labeled.

Contrary to what your academic courses showed you, datasets do not come neatly packaged with all the data labels that you need and ready to go. A big part of your overall data preparation activities will center around labeling your data.

This is another stage where involving your stakeholders will be critical. Not only do they know the data best, but this stage also presents you an opportunity to start building the relationships and systems that you will need later, during the human validation phase, after your algorithm has produced some results.

The key in this stage is to allow for plenty of time to label your data, you do not want to rush this process. Your project schedule will be reliant on time from your stakeholders and that's always a delicate proposition. No one on the stakeholder side makes data labeling a priority, so you will be getting small chunks of production out of your labeling efforts. The following are some best practices that have worked for me in the past.

1. **Labeling Team:** Within your stakeholder team, have them pick a labeling team that will be assigned to work with you on this effort. Pick carefully, this will most likely also become your human validation team in the future.

2. **You Need 3 People:** You're not always going to be successful at this, but a proper labeling effort needs *at least* 3 people working independently from each other. You have them label the same records and then you use a majority-rules selection process (or unanimous, if you need to be strict about it) in the background to ensure that you are getting the consensus of labels for the data. This is one of the few ways to build a high level of confidence in the labeling records.

   But, as I said, I recognize that this is easier said than done. It will inevitably lead to the three people breaking up all the records and each one of them labeling "their fair share" and handing it over to you. Naturally, the potential for both bias and error is high in this approach. However, you have to recognize when this is the only option available (or 3 people aren't available) and do what you can with what you have. Just make sure you set these expectations - and the downstream effect on your algorithm - with your stakeholders.

3. **Have Your Systems In Place:** How is your team going to label the data? What does the interface look like? How will you get the data back into your modeling systems? Where are you keeping your gold master datasets? You want to answer all of these questions so that your data labelers have a smooth experience. The better you can make that

experience, the more data they will (potentially) label for you and provide a better work climate all around.

Finally, I think most data science education programs do a pretty good job of introducing you to both supervised and unsupervised projects. Of course, with data labeling, I am referring mostly to supervised algorithms.

But you should also be prepared to work with unsupervised methods in the real world. The biggest pitfall around pulling data from documents is that the stakeholder will always tell you that their template for collecting data hasn't changed in years and that each document is the same as the next.

This is where your verification will pay off since it is rarely the case that you can write a single script to collect unstructured data. Take your time with these conversations and verify all the information you take in. You will likely find that collecting unstructured data can be done, but it is much more time-consuming and will require more code from your team to complete in an effective way.

A Word On Creativity

If some of the items around data preparation seem a little formulaic to you, you're not too far off. A lot of the tasks are going to be fairly simple with chasing down data sources, writing some queries, and then creating code to support those queries. Those parts are not going to be difficult at all.

But if everything was formulaic, then everyone across the globe would be doing the same things and there wouldn't be any room for creativity, right? That is far from the case when it comes to these tasks. I believe that there is plenty of room for you to be creative, perhaps more so than other phases of the data science development process.

When most people think about creativity in data science, they are mostly referring to the visualization phases. And visualizations are indeed important but I think it's healthy for a data scientist to exercise their creativity in other parts of data science. You should also view your creativity as part of the skill - and differentiating factor - that you bring to your team and your individual data science practice.

Adding width to your data lets your mind run free of factors that affect the business use case you are trying to solve. There is no single way to label data, so coming up with those systems frees you up to develop your own approach. Every conversation with stakeholders about their data will be different and they'll be looking to you for new insights. These are all opportunities for you to be creative as a data scientist. Don't miss them!

You might be thinking to yourself "Well, I'm not a creative person" and that's OK. This will come as you deliver more projects and get more comfortable with how you manage raw data at this stage.

The important part to remember is that it will be very difficult for you to find bonafide new insights if you follow the same patterns as everyone else. The answers are not always within a single dataset, insights are not always in the first query. You will have to dig deeper and with different shovels! How deep you go and what you find is up to you and the skill you will develop over time.

CHAPTER SIX

Model Search

It is unlikely that you will join the data science workforce with significant experience in testing multiple algorithms against your data. That is something that is very rarely done in a course of studies. If anything, academic curriculums are a little too segmented in that way. A course in regression methods is great, but all of your projects will *only* be trying regression methods.

From the perspective of the scientific method, academia has not served you well in this regard. In the real world, identifying the tools and methods that you will use for your data is a big part of the battle and it won't be easy. There will be no instructor to tell you which model to use or to even narrow it down for you. You will have to do it all on your own.

Champion Model Search

A well-executed champion model search (CMS) will save you a lot of grief over the life of your career. Unfortunately, many teams choose to take shortcuts here.

In simple terms, a CMS aims to find the best-performing model, with optimized hyperparameters, for the data that is available for the business use case you are trying to solve.

Pretty simple, right? It is. There are just a couple of items in that statement that we need to delve into a little deeper.

To find the "best performing" model, we need to determine the performance metric that we are going to use for comparison. Maybe you want to measure by false positives? Or false-negatives? What about F1 scores? There's really no wrong answer here, it will vary according to the algorithmic problem you have on your hands (categorization, predictive, signal processing, etc.).

In most scenarios, the metric you will use will be determined by financial values. You can usually assign a monetary cost to metrics like false positives and false negatives. Your business stakeholders can help you with this and you should allow that to drive your decisions.

For the hyperparameters, you will have to involve a grid search to find the combination of values that serve your model best. This is the most time-consuming portion of a model since the number of permutations - assuming you are testing multiple hyperparameters - can add up quickly. Make sure you allow for plenty of time for these tests in your project time estimates.

Where the CMS really takes off is in the realization that there are many models to try in this process and you should take a "no stone unturned" approach to the models you will try.

You will be tempted to think about data science as having the "traditional" methods, like regression, and the more "advanced" methods like neural networks. But I would caution you that this is the edge of the slippery slope filled with bias. While I will provide more on bias later in this chapter, I have to start by advising you that there are *no* traditional or advanced approaches to data science. Models are not to be thought of as newer, older, better, or anything else along those lines.

You are a scientist, with a data problem, and are standing in front of a toolbox with many tools. Each algorithm is a tool that can be applied correctly (or incorrectly!) but you'll never know until you try them.

Yes, I just described a process that will take a long time. How many models you attempt with your data is ultimately up to you. You can make educated decisions on which models to try based on your data, but be careful to not go too far in your rationale for excluding models.

For a frame of reference, in my graduate studies, a capstone course had a requirement to attempt 13 models on the same data. On the data science teams I lead in corporate environments, we will attempt anywhere from 6-10 models for most projects.

Yes, that's a lot of experiments to keep track of. Yes, you might find yourself working late on this. But this step is a critical part of the methodology I laid out earlier in this book. This is all part of the diligent work that will let us sleep well at night.

It should be noted that there are some tools out there that greatly reduce the work that is required in a CMS. For example, some companies will resist a tool like Data Robot because of the way it handles the final deployment of algorithms.

But if you look at the earlier parts of the Data Robot workflow, it is an effective tool for trying many models at once and can provide you with (at least) some high level guidance on which tool fits your data best. While that's using only a portion of Data Robot's capabilities, it's a valid example for how a CMS process does not have to be onerous on your team.

Iterative Development Process

You will most likely have to go through a series of model versions to get your algorithm to a state where it can be used by your business. You simply won't get everything right the first time and that's OK! I put all of these items under the umbrella of the iterative development process. It is a workflow that you will follow many times throughout your career.

This is not something you are likely to have done a lot of in academia but the process is not difficult either. Below is a chart that illustrates the process.

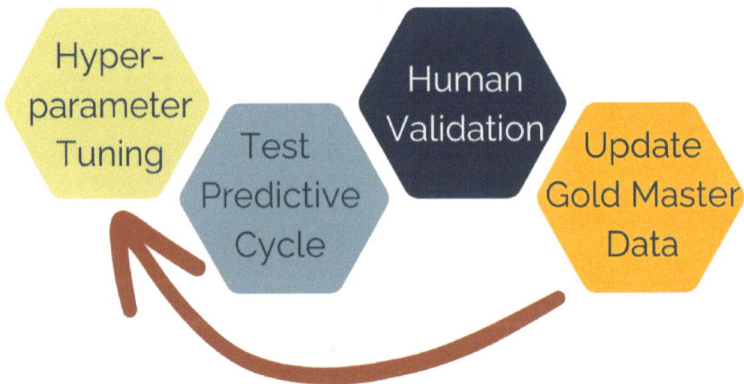

Recall the earlier chart that I used to illustrate the methodology of data science. That process contains a loop of sorts between **Model Design** and **Interpret Results**. You can consider the above chart the different stages you should be going through for that loop.

- A proper grid search for hyperparameter tuning is likely to be your largest time investment in the whole process and you should plan accordingly. It is important to not rush this step since you will most likely need to cycle through a large number of permutations to find the right hyperparameters for your data.

- You should aim to collect the stakeholder feedback as efficiently as possible. The format should be easy for them to fill out and easy for you to feed back into the model. While I present more information on human validation in other parts of this book, you should know this step won't always come easily or quickly and you should be prepared for it to affect your project timelines.

- The process that is illustrated here is not just for use during the initial development of the model. It is also the basis for future maintenance of the model as new data becomes available and/or the needs of the business change.

The finish line and deployment date of your model can become opaque during this process. How and when that finish line is reached will be a combination of the traditional metrics from your model and the feedback from your stakeholders. Tangental to that will also be the time & resource constraint that any project has. What you can negotiate during this process will determine when you can stop the loop and move into other parts of the development process.

Always keep in mind that the algorithms - and the way you think about them - can't always be focused on code & math. The business wants what the business wants.

A Word On Bias

Recall my previous assertion that data science, in its purest form, is the removal of bias from your decisions. If you can do that, you are making data-driven decisions to effectively guide your business.

If there is one step in the entire model development lifecycle that will allow bias to creep into your process more than the others, the model search process is it. I have seen too many teams and too many projects implement poor model selection processes that have downstream effects on the ultimate success of your algorithm. Here are some examples that I have seen from my work in the field:

- **"Code Snippet Sniper":** This is when a team selects a model "because we had the code snippets from another model . . ." Just because the data for two models looks to be about the same does not mean that you should just carry the code over and re-use it. You have to treat every single dataset and project as being unique and you can't assume that the same code will give you the desired results. You are also making an error-by-omission because you are not trying models that could be *better* than your current results.

  Another reason I think acknowledging this bias is important is because it can also affect the time requirements you create for your project. If you think the data is the same, and you plan on using the same code snippets, then you will also be tempted to think that the project will take a similar amount of time and you may give a bad estimate to your stakeholders. That's a dangerous situation - for everyone involved - that can be avoided with a model search process that begins with a blank slate every time.

- **"Rusty But Trusty":** It seems like every experienced golfer out there has that one club in their bag that they always carry and use. It's old, worn down, and rusty, but they just seem to always be able to hit the ball well with it. That club is known as the "rusty but trusty", it just never lets you down.

  How did you pick your last model? Was it truly the best model or is it just the one you feel most comfortable with? Do you have a favorite model?

You shouldn't, that's a form of bias. There is no favorite model. No two data projects are the same and you need to conduct a proper model search every time.

- **"Whitepaper Warrior":** It's fun to read whitepapers, they have their place in your growth as a data scientist. But it's a massive mistake to read a white paper and then implement it in your next project. You run the risk of the theory presented in the paper being wrong. Or maybe you didn't understand the approach & requirements correctly. Or there's an implementation risk with you translating a paper into code.

  Read white papers, but let them sit in your mind for a little bit. This allows you to process the information while also letting further research (by you and global groups) decide if the approach is valid or not. In other cases, you might need the approach to be adopted into a library. Resist the temptation to start a model project by saying, "I just read a white paper that addresses this . . .". Take your time with that research and implementation.

- **"Bleeding Edge Billy":** Another part of the fun of being a data scientist is seeing all the new cool tools & libraries that are out there. The marketplace for these tools is certainly dynamic with both new tools and significant updates to existing ones. I like to tell my teams that we need to take a balanced approach to this. It's a mistake to ignore the new developments & tools but it's also an equal mistake to chase "the latest & greatest". You shouldn't aim to solve a problem with a new tool just because it's a recent release. You should use whatever is the right tool for the job. So, just as with white papers, resist the temptation to start a model project by saying, "I just downloaded this new tool that addresses this . . .". Take your time with that research and implementation.

A Word On Notebooks

As you grow more & more in your career, and the complexity of what you work on increases, you should find yourself using tools like Jupyter Notebooks less & less. I fully understand that this may not be a popular opinion, but I stand by the old saying, "You can't deploy a notebook". I don't make the rules, I just report them!

When you increase your use of Python in a professional environment, I see a few things happening:

1. You will start to see Python as less of a scripting language and more of an object-oriented programming language that you will use to write *programs*, not scripts.

2. You will need to do more complex coding in Python with tasks like multi-threading.

3. You will find yourself working in more complex, distributed, server environments with advanced deployment needs.

All of these items are very, very difficult to accomplish within the confines of a notebook. This points towards you working more with Python from a traditional integrated development environment (IDE) and executing from a command line interface (CLI). The earlier you adopt this approach, the easier the transition will be for you.

CHAPTER SEVEN

Model Deployment

While I don't claim to be an expert in every single data science curriculum that is out there, I have yet to see any given course, at any academic level, that spends much time (in most cases, any at all) teaching students how to deploy models. You create the model for a class, you get the right answer for the assignment - most often just running on your local machine and not a server - and then you put the model away, never to be seen again. This is a topic that courses simply don't invest any time in and will be your weakest link when you are in the first year of your data science career.

Model deployment can be straightforward and complex at the same time; entire books could be written on this topic. I will try to present as many high-level concepts as I can in this section and it will be up to you to decide which topics you want to dig deeper into, based on what you need for your data science practice.

If you walk away from this section feeling like you learned more about traditional software development rather than data science, then that's the right idea. I always tell my teams that in the late stages, data science projects end up converging more & more with traditional software engineering deployments. Your models should end up resembling libraries - just like any other programming libraries - that you are deploying to your user base.

Prediction Runtime

Sure, maybe you had one project in your final year of studies that gave you a modest dataset to make predictions against. You did that once, but probably haven't done it since. However, being able to think about and design your models in a way that will allow them to make predictions in your professional work will require significantly more effort. Your primary needs will be:

1. **Scale:** In many real-world applications, you will be working on a much larger scale than what you could have possibly done in academia. You will need to be ready to run your models through vast data sets.

2. **Performance:** In what type of environment will your model be used? What are the performance metrics required for each prediction? These values can vary quite a bit depending on the type of algorithm you are working on. In a lot of cases, these requirements will be dictated to you and you will have to make sure that your model performs accordingly.

3. **Repeatability:** Unlike academia, in the real world, you will rarely work on a model you only use one time. Most models rely on long-term use to create the return on investment (ROI) to keep you & your fellow data scientists employed. You will need to create both models and runtime prediction processes that can stand up through repetitive use and give consistent answers.

The best way that I have found to achieve these goals is to think about your model as a single component within a greater program that you are trying to deploy. Said another way, your model will eventually be consumed by other pieces of code that will handle the communications between the model and whatever other code needs to access it.

Let's think about the following example from the world of natural language processing (NLP). We will pretend that we are working on a classification task that takes a piece of text and classifies it into one of three categories. In broad strokes, your model may consist of something like the following:

1. A way to gather the text input

2. Pre-processing for stop words and stemming

3. An approach for tokenization and maybe managing your dictionary a little bit

4. A mechanism to send your tokens to your model

5. A numerical return from your model for the categories

6. An interpretation of the numerical return and display of your category

I think this collection is a fair representation of what you may have done with NLP in the past for your classes and I'm guessing you've spread this work across multiple code files.

To achieve both the scale & repeatability that we need, your challenge is to take all of these components and create a single library from them. Here's the basic framework for what you will need for that:

1. A library with a single method that takes in 1 piece of text. The return value from this method is the name of the category that your algorithm has allocated for the text value. Why just a single method? This is where the scalability comes in. You want to create the most straightforward method you can so that it can be called as often as

needed (from outside your program) and be used at the required scale. Yes, you could be doing that same scale from *within* this program, but you will set up your future implementation best if you keep the libraries at this level as simple as possible. There are more complexities to come in the code that will consume these libraries.

2. A single class that handles all of the text pre-processing that is required. This class will also need to be supported by other things such as calls to your dictionary methods, when needed. Also, realize that I am only using NLP as an example here. In other cases, you may need to pre-process your numerical input (scaling, etc.) but you should still have a dedicated section of code for this.

   Please resist the temptation to do this outside of this class. You *could* be doing some of this pre-processing before even calling this method, requiring your GUI code to already send in processed data. But that's usually a bad idea. Not only are you creating the potential for mistakes, but you are also creating a maintenance nightmare down the road.

3. A single class that only contains the prediction method for your algorithm. While this may seem like a basic thing to point out, as data scientists, we usually end up mixing our training code and prediction code within the same files and/or classes. You will usually need to do some re-factoring of code to break the prediction code out in an appropriate way.

If you can find yourself implementing a library in this manner, then you will be ready for the next steps.

Application Development Layers

Modern applications are consumed in any number of ways. Your algorithms might be used to power mobile apps, back-end server systems, web apps, or any number of things powered by a variety of languages. You have your Python library as your core asset - and that's good - but now you need to wrap your next layer around it so that it can be used by different applications. This is when you will most likely need an Application Programming Interface layer or API.

When you create an API, your goal should be to expose a single endpoint that can then be accessed from the outside world. That endpoint will serve as the border between you and all the other programs that want to use your algorithms. What those programs are is only of passing concern to you. And how you produce the data that feeds the endpoint (i.e. your algorithm) is of no concern to the other developers. Most modern development is handled via Representational State Transfer, or REST, APIs.

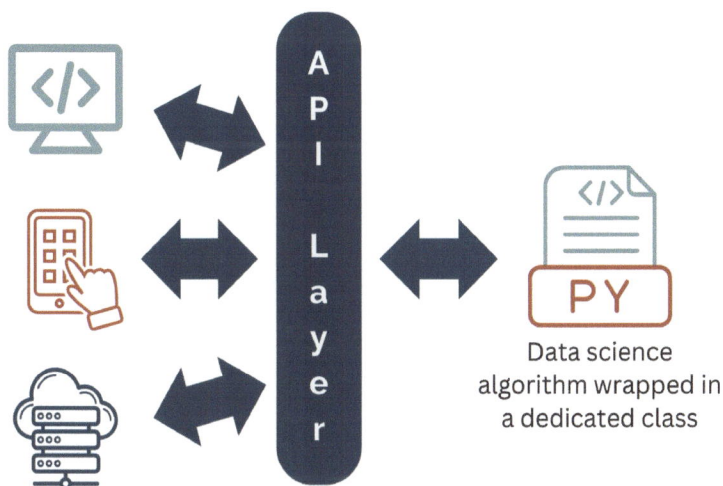

Data science algorithm wrapped in a dedicated class

The beauty of working in this manner is that REST APIs are essentially a universal protocol that modern programming languages and systems adhere to. They can be created & used by any number of languages and that can make them quite fun to work with!

In one of the last C# projects that I worked on, I was the front-end developer for a cryptocurrency startup. It was one of those real "go, go, go!" types of environments where we all sat around a traditional scrum table and worked hard. I was paired to work with one of the co-founders since he was the primary back-end developer. We would sit there for hours and just throw API endpoints back and forth at each other. I would tell him when I was ready for another piece of functionality and he would tell me the name of the endpoint after he quickly created it on his end. I would then write the code to access it and see if it worked.

He would sometimes send me real data, and other times dummy data, but it didn't matter because the important part was the connection between the two systems; you can always feed real data to an endpoint later. The best part was that I was working in C# and he was working in NodeJS. But it didn't matter, the REST API layer takes care of the work in between. We would sit there for hours just throwing code around the table, it was fun!

I can't talk to you about API layers without also discussing security. Up to this point, all of the code that you will write for data science you can probably handle within your team. But when it comes to exposing endpoints and letting other applications use your algorithms, you also need to implement a security plan. At the very least you will need to manage state (or user sessions) for every inquiry. You will also need a plan for authentication and authorization for each of those sessions. You likely will have to work with an IT/IS team within your organization to finish out these API layers in a safe manner.

I completely acknowledge that this section could have been included in any number of technical books. These concepts are not unique to data science. I want to emphasize again that you will wear many hats within a data science team. Teams are small, they will require you to be flexible and move with the needs of your project.

You may find yourself writing these application layers simply because there is no one else available to do them. Besides, I think it's healthy for you to use your programming skills - particularly within Python - to do something else besides data science. Python is an extremely capable language and extending your skills into this type of coding should bring that to light a little more for you.

Graphical User Interfaces

In most of your data science studies, you were probably the only user of the models you developed. Making data science your profession will be a polar shift; if you are the only one using your models, then your model doesn't exist.

We've discussed the programming side of model deployment, but you will also have to create the visual side of these apps to make them appealing to your users. Graphical User Interfaces (GUIs), and their development, are a skill you will have to develop.

And, no, Jupyter Notebooks are not a GUI!

There are many excellent books & resources already out there for app design, I don't intend to re-hash them here. However, I do want to reinforce the simple approach that I've presented in this chapter.

- You will want to extend your programming skills to include some HTML and the basics of web design. It will also help to know the basics behind AJAX and how you can use that for dynamic sections of code within your GUIs. This doesn't have to be complex, there are plenty of toolkits - like Bootstrap or jQuery - that can help you with this.

- You will also want to know a Python web framework like Django or Flask.

- There is no one way to learn about creating your proprietary APIs. The best way to learn about API design is to work with as many APIs as you can and understand the different tasks each endpoint addresses and how you can fit that to your own needs.

You should also know that GUIs are not just for your deployed models. They can also help during your model development process.

1. Your supervised models may require labeled data that does not exist yet (or not enough of it). Developing a GUI where your domain experts can label data for you is an excellent way to involve your stakeholders as well as streamline your data collection process.

2. When your model produces results, you will again need your domain experts to validate those results. While there are many other ways to collect this feedback, creating a dedicated mini app where your stakeholders can do their work can both be fun and add a little bit more professional "polish" to the work your data science team delivers.

These two scenarios are also a good way to introduce some gamification into your enterprise. Labeling or validating data is a thankless task that simply needs to be done. By creating an engaging app, you will be able to both collect data and make things easier for your stakeholders.

In one case, I always noted that our summer interns had a lot of downtime without much work to do. This is particularly true at the beginning of the summer when their respective managers were still trying to figure out what to assign them. So I would develop quick GUI interfaces to have them help me with tasks like labeling data for my NLP algorithms or some basic categorization tasks. I would then award a company pin to the intern who had the most labeled entries for the week.

In another case, I needed a lot more labeled text data and I ran a department-wide effort for 6 weeks. This required a much more complex web app to determine which data to show to particular users as well as track scoring for both the individual and the sub-team they belonged to. I gave prizes for both top individuals and top teams and that motivated the entire department to help us complete our labeling on time.

It will be up to you to figure out how gamification fits best in your organization. But there's little doubt that it can help you in some way and another reason why GUI development will be important to you during your first year in data science.

Putting It All Together

If you want to put all of this together, here's an exercise that you can walk through. Just as before, you can choose to use this as a thought exercise or you can decide to code this out. Either way, I think this will prepare you for some of the same tasks that you will be given during your first year in data science.

We are going to create a small game where you will ask a player to guess the temperature on a given day. They will be shown the predicted temperature and will have to guess what the actual temperature was on that day.

1. Create a class that handles some basic gameplay functionality like:

1. Pull a new entry from the weather dataset that was created from an earlier exercise in this book. You should collect both the predicted and actual temperature values.

2. Adding a +1 value to a player's score. This class doesn't have to be fancy, just something that we can access from another layer.

2. Create a new script that will serve as your API layer. You can follow a tutorial for Python Flask since it's lightweight and easy to implement. This script should:

   1. Expose an API endpoint for a user to enter a value.

   2. Maintain session state for a user. You can choose to manage the user score within the session. This also allows you to have multiple users at the same time.

   3. Start up an HTTP server that will host your API on a unique port number and serve up your API to the outside world.

3. Create a Flask HTML template for the gameplay. Again, it doesn't need to be fancy, we just need something to tie the entire exercise together. The template just needs to present the user with a new weather forecast and collect their answer on the actual result. Then display a new score based on whether the user was right or wrong.

Still with me? By now, you have taken some code from a class, wrapped an API around it, exposed an API endpoint, and put a GUI on it so that it can be accessed by a layperson. If you decide to build all of these components, then you have successfully taught yourself some of the same concepts that you will need in the real world of data science. Great work!

Model Portability

Another consideration for your development process will be the portability of your model. Throughout the entire lifecycle, your model (and its components) will be on a variety of systems. Maybe your initial development is on your local machine and within a Notebook. Maybe the initial data from the stakeholder was in an Excel file.

But now the model & data have grown and you have moved it to a development server. And maybe you already know that it will need to be moved again onto another server for the deployment to production. That's a lot of growth and movement for any code to go through!

You will have to make sure that you have testing procedures in place to account for any operating system and Python library dependencies as well as for any performance metrics that you want to achieve. As presented in earlier sections of this book, if your model requires fast response times for predictions, then you will need to make sure that you can simulate and test this for any environment that your model will come across.

You will also need to coordinate with IT/IS teams to make sure the Python environment has the libraries you need since these servers may not be under your direct control. All of these are items that you will address in every model deployment you do, not just in your first year.

Model Maintenance

Earlier in this book, I presented a chart for the iterative process a model goes through to be developed. Maintaining a model will follow a similar pattern.

You are welcome to develop your own, but the following is a pattern that has worked for my teams in the past.

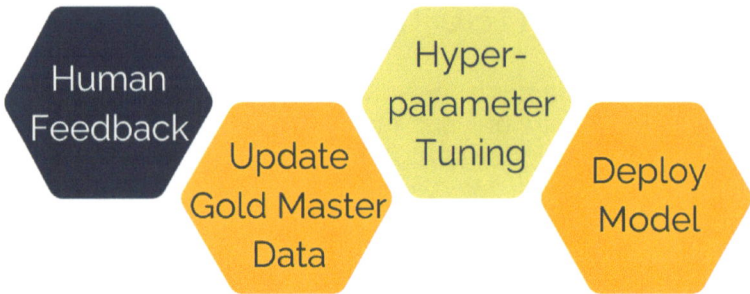

The frequency of model maintenance will also be driven by your stakeholders and the nature of the model. Some models - like a pricing model - may not need to be updated for years. Other models - like some image recognition models - may need to be updated very frequently since images tend to be highly dynamic datasets to work with. You serve your stakeholders best by observing and reporting the performance metrics of the model to your business and allowing them to make an informed decision. Make sure you have these monitoring systems in place prior to deployment. You should be particularly alert to any significant degradation in these metrics and report them accordingly.

It should be noted that some of this maintenance might also trigger other policies within your organization, such as model review & approval requirements. It's important to know these timelines ahead of time so you can set these expectations with your stakeholders since they can have a tangible impact on your delivery.

The Cloud

One of the most popular questions I receive from my social media followers is for my thoughts on cloud certifications. Most people find themselves having to choose between one of the popular platforms, like AWS or Azure for their studies.

Generally speaking, you will find a fair balance in the market for employers that use each of the platforms. So that means there is no inherent advantage to choosing one over the other. There is no strategy here, so don't overthink this. Just pick one and finish out the program to earn the certification.

But you should also consider that there are plenty of employers - even entire industries - that do not use a cloud approach. Or at least not a "cloud first" approach. There are plenty of companies out there that work solely with on-premise solutions in their well-developed server centers and have little to no interest in your cloud certifications.

You should balance out your skills by making sure you can work directly with servers (and clusters) from a command line interface. You don't have to be a genius with UNIX/ Linux, but you should know enough to manage Python instances and execute your own programs with little assistance from a GUI.

Other Languages

This seems like a good time to reinforce an earlier concept. I hope that this discussion has shown you why Python is just so dominant in the world of data science, especially when compared to R or SAS. It's these deployment activities - the "last mile" of data science - where R & SAS fail to deliver a scalable and consistent solution for end users.

Generally speaking, you can do all of the math & programming for data science tasks in all three languages. But when you start needing the API layers, the GUIs, and the required speed for fast processing, Python is the only language of the three that can handle that.

CHAPTER EIGHT

Stakeholder Management

If I were to choose one of the topics in this book as being the most challenging, stakeholder management might very well be it. No two situations will be alike and there will never be a single correct answer for the questions you will face. There is also no finish line here; I am many years into my data science career and I am *still* developing my stakeholder management skills.

Don't despair - the greater the challenge, the greater the reward. How you properly manage your stakeholders (or don't!) will have a massive impact on the success of your projects and carry on downstream into your career. Stakeholder management can be challenging, opaque, and downright frustrating. But it can also elevate you as the point person for your team when other people need something and that's a valuable position to be in.

Communication

When you are coming out of a data science study program, you've convinced yourself that the world shares the same vocabulary and understanding of what algorithms are and what they are capable of. Surely, your stakeholders already know what's going on in the world and the greatness that data science can do for them. Right?

Wrong.

Your stakeholders are primarily going to be business professionals, more likely to be aware of the latest marketing & accounting terms than data science. And they almost certainly won't know the math behind the algorithms you use.

Quite frankly, they are not going to care, nor should they. They will always care more about the problems you can solve for them and the accuracy of the solution you are proposing. How those solutions are created will be of limited interest to them. The algorithm you create for them can be some complex math or it can be a room full of monkeys that you have trained to use iPads; the stakeholder just wants reliable results that carry significant business impact.

You have to meet people at their level. If you agree that you will be working with stakeholders that want business impact from you, then you need to speak in those terms. You need to avoid math & code at all costs in these conversations. Presenting formulas and data science theory to stakeholders is, most likely, a mistake. I have used the following framework to assist with a straightforward approach to speaking with stakeholders:

1. Here is the business problem we want to solve
   "We want to increase sales by finding an algorithm that better identifies potential clients"

2. Here's how many approaches we tried
   "We experimented with seven different algorithms and 3 years of sales data"

3. Here's the approach that worked best

"A neural network-based approach gives us the best results". That's it - no need to give a more technical explanation than that.

4. This is the impact this approach can have when deployed.
   "Using last year's sales as a baseline, our research shows we can increase sales by 10% this year"

If this framework seems simple, that's because it is. The whole point here is to not just be a data science team but to be a team that is focused on business solutions.

If all you do is watch YouTube tutorials and data science "influencers", you might have been convinced that all data scientists do is talk in math and code. I fully understand that this may be the first time you've seen a framework laid out in this manner. But, trust me on this, if you want to have good meetings with your stakeholders, leave the code talk to the data science team room.

Education

If you take ten business people and ask them something like, "What is machine learning?", or, "What is data science?", you are going to get ten completely different answers. As practitioners of data science, you & I are comfortable with these questions. However, for the general public, there is still a lack of both consensus and understanding of many concepts within data science.

Remember how I said that you have to meet stakeholders at their level? It will be up to you to provide some education to your stakeholders on what exactly you are talking about and how that concept relates to what they are doing or asking of you.

What does this look like in practice? You might find yourself having to lead some internal seminars, like a Lunch & Learn, where you can (slowly) introduce data science concepts to your colleagues. The goal here is not to teach them data science. You shouldn't even show any math or code. The greater point you want to get across is the following:

> *"Our company just put together a data science team and I am part of that team. Here is what we're capable of and here's what we're not capable of.*

By listing what your team can and can't do, you are opening the door to have people come and talk to you to better understand how they can utilize you.

Be ready with an icebreaker that is also a data science activity, and be prepared to show examples of good projects and bad projects. Give them an idea of the role they will play (labeling, human validation, etc.), and make yourself available for questions. All of this will make future interactions with your stakeholders a lot easier.

Ideation

For many of you reading this book, this will be your first job in data science. For a significant portion of you, this may be your first job ever. If either one (or both!) of these apply to you, then this is where things get a little more difficult.

This book also contains a discussion on model selection and how your academic courses essentially pre-determine which models to use. Another pre-determined aspect is what you're going to do with your data and the model. Your professor and/or the curriculum take your creativity out of the process and tell you what information to pull from the data. There are very few (if any) opportunities for you to work with the data on your own, come up with your ideas for assignments, and solve them on your own. I wish there was a way to prevent this in academia, but there is no method I can think of.

In your first job, you will need to be the person who comes up with the ideas of what to do with the available data. Both the ideas and the models will not be handed to you via a class assignment. The stakeholders around you will be looking to you to provide ideas of what is possible with data science. That can be a lot of pressure!

You will need to find ways to learn about your stakeholders and their business quickly. You need to be able to understand their workflow processes as well as their pain points that you can automate. Integrating yourself into their team - perhaps through a shadowing effort - might be the quickest way for you to learn more about them in a matter of a few days.

It will be up to you to create these partnerships and find those opportunities to spend time with the stakeholders and observe and take notes. That should get you started on a few ideas that you can present back to the stakeholders in the future.

If shadowing is not possible, then there is a better solution. The ideal solution is for the stakeholder to present you with data science ideas that they have identified and that you can explore with them to see if there is a mutually agreeable project there. This will be made much easier if you have already taken the time to educate your stakeholders and they are clear on the projects your team can deliver.

If that's not the case, here is a shortcut that I have used with some of my stakeholders. I will ask stakeholders to think about all their pain points or all of the things they wish they could do more effectively or anything else they think is a data science problem. From there, I tell them:

> *"I want you to think about your idea in your mind. And then I want you to tell me if you can solve it using Excel. Maybe you don't have all the data or maybe you don't know all the formulas. But let's assume that I can get you both the data and the formulas, can you see yourself reaching a solution for your problem within Excel?"*

And then you wait for an answer. Give them plenty of time.

While there may be exceptions to this, my position is that if an idea can be solved in Excel, then it is most likely *not* a viable data science idea. It's most likely just some form of advanced analytics or applied statistics.

But when a stakeholder comes back to you and says they have an idea that *can't* be solved in Excel, that's when you should pay closer attention because you might just have a data science project on your hands.

Data Security & Discretion

This may be a rather obvious discussion to someone with a few years already in the field, but I want to make sure that I also throw in a quick word here about discretion. There are many companies out there that are in a growth phase for data science usage and are throwing more & more data at algorithms. They are also more willing to push the edge in the algorithms that they deploy.

Through that process, you may be exposed to different datasets that are otherwise unavailable to other teams in the company for any number of reasons. Whether that's financial data, customer data, employee data, etc. you will need to exercise your discretion when handling this data and how you speak about your projects. The less you say and the more security you put around your datasets, the better. You do not want to erode the trust of your stakeholders if some data were to be released in an unintended manner.

As an aside, from a data science executive perspective, I stress the need for data security to all members of my team. I have worked for organizations that go through a formal goal-setting process each year. And I usually create a goal around data security for my team. It will be something along the lines of "My data science team will not be responsible for any data breaches for any of the projects that we work on; we will exercise stringent data security policies".

I share that with you because your actions around data security reflect on your manager and your entire team. By doing your part, you are helping maintain the reputation of the entire team while also taking a professional attitude toward your data science career.

Human Validation

In your academic studies, you most likely followed a pattern that had you finished with a model after you generated the right answer. That model was then put away and never used again. But, as you've seen in other sections of this book, your first year will teach you about the iterative development process that a model must go through to reach a deployment stage.

One of the most important parts of that iterative process will be the human feedback that you get from your stakeholders. This will, most likely, become one of the times when you have the most contact with your stakeholders since their feedback is so vital to you. In your data science studies, *you* were on the human validation side; your algorithm either matched the prescribed output or it did not. You will not be able to do the same in the real world.

There are a couple of phases that you will have to go through to have a good relationship with your stakeholders and carry out the human validation requirements. I have tried to lay them out as best as I can, but these approaches will vary based on the size of your team, the specific project requirements, the scope of the model, and any other factors. I will leave it to you to adjust accordingly.

First, accept that you & your data science team do not have all the answers. You are a smart & talented team, but you are not a business team. No one will know the data and the results better than your business stakeholders, so resist the temptation to proceed without their feedback. This is one of those "stay in your lane" scenarios where you need to know what your team can and *can't* speak intelligently on. This is a skill that most good scientists (not just data scientists) possess and will serve you well in the future.

Related to that point, involve your stakeholders early & often in your model development efforts. Your stakeholders need to be involved in all decisions on both the design and the output of the model since they know the business side best. Not only are you creating a good bond with your colleagues who will now be learning more about you & data science, but you are also planting seeds for future requests. You need to have your stakeholders feel like they are invested in the success of the algorithm because you are going to make harder & harder requests in the future.

Speaking of those hard requests, you will need to ask your stakeholders for human validation of the results that your algorithm produces. This becomes a little harder because you will request both more people and more time from your already busy stakeholders. If you've involved them enough in the model development process up to this point, then this request will become much easier. Here's how you can make this process more palatable for both you and your stakeholders:

1. Have a specific ask for both what you need to be validated and how many records you need.
   "We need your team to validate 1,000 results from the algorithm. We estimate that it will take a 3-person team 2 weeks to complete it"

2. Prepare a delivery format that makes it easy both for your stakeholder to fill out and for you to ingest back into your model's training data.
   "Here is a mini web app that we have created where you can quickly look at the results and provide your feedback"

As an aside, here is something else that you probably didn't have to do in school. At some point, you will most likely be working with data of varying age and quality and, in some cases, maybe even some synthetic data. All of that will be enriched with the human validation we have discussed in this section. This is where you will benefit from maintaining your **gold master** of training data. That is a common practice where you only put your newest, human-validated data into this set, and it's the primary one you use for model training.

Protect the quality of this data set at all costs. You're going to want to segment your training data in this manner because you, most likely, have never needed multiple sets of training data before.

Demonstrating Impact

No professors are handing out grades in the real world. You will have to adjust to a completely new way of both looking at & evaluating your work to determine your own markers of success. One of the most common markers is the business impact that your project brought to the proverbial table.

Recall the earlier discussion on communication with your stakeholders. Demonstrating impact benefits from your communication skills because you will have to create accurate representations in plain words. The *last* thing you should be doing is thinking about code & math at this stage. This is where you need to be a visual storyteller while also proving results in terms that are meaningful to the business (not just your data science team).

Each project (and enterprise) may have its own defined measurements and you should follow those. Here, I present some of the common markers that I have come across in my career.

- **Work Hours:** Can you assign a particular number of work hours to your algorithm? Did you create some sort of automated process that can now save the company human hours?

  This is most common for machine learning algorithms, particularly categorization tasks, where you have a baseline of human hours that you can compare against.

- **Scale:** Did you create an algorithm that can expand the amount of data that you can now process? Or does your algorithm now consider many more factors than any human could do on their own? Good. You can demonstrate impact through the scale of your approach.

  An example here could be a neural network project that you are using to replace a manual process. Maybe a sales agent can only take into account 5 user attributes and process 10 sales prospects each day, but your neural network can handle 100 attributes and process thousands of prospects per day.

- **Near Misses:** Does your algorithm identify and/or prevent some sort of financial loss? Or maybe identify some sort of regulatory risk? Great, that's a form of impact too. You can usually assign some direct dollar value based on what your algorithm identifies and the transactions involved.

Maybe your algorithm identifies fraudulent transactions that were caught early and saved your company a particular dollar amount.

Note that this whole process should mimic a fairly straightforward analytic exercise for you. In this case, the root dataset is your results from the algorithm. Then you take that data (or as much as you have of it) and apply it to one of the markers above with analytics. That should then give you a dataset that demonstrates the impact of your project and that is what you use to build the visual story for the business.

Finally, keep in mind that this phase will also introduce the return on investment (ROI) for your algorithmic projects. You should be open to the fact that your algorithm may be interesting and effective but does not produce a justifiable ROI. Maybe your algorithm is really good at identifying fraudulent transactions but there's just not enough of those transactions (or a high enough dollar value) to cover the costs of the algorithm.

Ah, yes, costs. We can't have a discussion on ROI without considering at least some of the costs involved with an algorithm. I'll save you from the obvious costs like your salary, GPU processing hours, etc. But I will point out one of the costs that may not be obvious to you. Algorithms, when deployed properly, tend to stay active for a long time which means that your team will need to not just develop but also support the algorithm for a few years to come.

This creates a form of overhead for every algorithmic project that you are evaluating at the proposal stage. Algorithms will have new data come to light, performance might degrade over time, new hardware will become available, new policies will require model reviews and/or additional documentation from you, etc. There are many things that you & your team will have to do to keep an algorithm going and all of these tasks are likely to have a material effect on your ROI calculations.

It's possible that some of this discussion will not directly apply to you during your first year; your data science leadership may address some of these for you. But it's still good for you to have an understanding of this as you sit in stakeholder meetings and start to evaluate your own data science project ideas.

CHAPTER NINE

Data Science Team Dynamics

Not all aspects of your first year in data science are going to be as tough as some of the previous sections. Some of them can be quite fun and integrating yourself into a data science team - and the collaboration that comes from that - is one of the great aspects of your first year.

Just like everything else in this book, it will be a process that you will have to experience to learn properly. But with the right mindset, working with other data scientists (of varying skills) will help you both personally and professionally.

It should be noted that there is no single way to run a data science team (or any technical team for that matter). So, I can only share my experience in running my teams and this approach has worked well in the past.

In this chapter, I lay out the processes as I see them and I will leave it up to you to adjust for both how your particular team runs and the ways that you are personally comfortable participating.

Team Meetings

There are two primary types of meetings that I run with my teams:

1. **Technical Touchpoint:** A weekly meeting where we can discuss any of the items that have occurred in the previous week. I consistently advertise that I am the most senior person invited to this meeting, none of my supervisors are allowed to attend and I don't share my notes with them. This frees up my team to "talk shop" and feel like there are no stupid questions.

   I try to avoid anxiety in this meeting by staying ahead of what everyone on my team is working on. No one is directly called upon in this meeting. It is up to me - the manager - to know & recite what everyone is working on and then they can correct me if my understanding is not right.

   In some cases, I may ask a team member to demo a particular model they are working on or someone may decide to present some interesting articles they are reading. In other cases, we might talk about a problem with a model or otherwise collaborate on a solution that is not focused on blame, but rather on getting to the right answer.

2. **One-on-One:** A monthly meeting where I meet individually will all my data scientists. This meeting has no agenda, I let them decide what they would like to talk about and it doesn't all have to be about work. This leads to open conversations about work, family, personal concerns, and anything either one of us wants to talk about. It helps create a stronger team and one of the ways that I lead with empathy because I realize that everyone has lives outside of work that still need to be talked about.

Will your first year be like this on your team? Maybe or maybe not. But I can still expose you to this framework to perhaps enact change in your organization or start to form your approach to management as you grow in your skillset and eventually become a data science leader in your own right.

Project Management

Another transition that you will have to make is the adjustments to deadlines that both your stakeholders and superiors will hold you to. This should be helped by whatever project management system is in place on your team. This also might be the first time that you are working under a system like this.

I could tell you about Agile planning, waterfall planning, ticket management, and all kinds of things like that. But there are entire books - even series of books - that can already help you with that. It behooves you to educate yourself on the systems that are out there.

For now, I prefer to present you with high-level guidance that I give to my data scientists and you will have to find ways to integrate them into whatever project management process your team has. The following is simple advice but I think it will serve you well no matter where you end up working as a data scientist.

1. **Two-week work cycles:** If you've heard of Agile planning before, then you might be accustomed to hearing about "sprints". I have deliberately avoided using that word in this description because it may or may not apply to your team.

   However, there is a reason why sprints exist. The rule from Agile is that if a task cannot be completed within a sprint, then it needs to be broken down further, it simply hasn't been defined enough. I think that's pretty good advice to live by and endorse that theory, no matter how you go about planning your project.

In preparation for your future work in the field, become accustomed to thinking about what you can accomplish in two weeks. It will require some experimentation on your part to know exactly how much work & code you can produce in that time.

Use that to help drive the types of tasks you can (and can't) accomplish in that time and how you would express those tasks in team meetings. Later in your career, you can also use this as a measuring stick since you should be able to accomplish more & more in the same period as you develop your skills further.

2. **No bad dates:** Realize that project management is a big part of being a data science manager. More importantly, your manager will make promises to others based on conversations with their team. You help your manager best by giving the best time estimates that you can and not providing an unrealistic date.

   If you don't think you can deliver a task by a particular date, please speak up and say so. While there may be a slight feeling of embarrassment or anxiety when you say that, it's the better choice over giving a bad date that you are likely to miss. When you are having more direct contact with stakeholders, you should carry over this rule into the conversations with them too. You'd be surprised how much more your colleagues will appreciate your honesty in your time estimates. And you are not working as slowly as you think, just because you give different dates.

Remote vs. In-Office

By way of my social media content and live streams, I usually get some variation on the question of remote data science roles versus in-office roles. I can understand why that has become a key question in modern times.

Brace yourself, I'm about to take an unpopular opinion. I love the convenience of a remote role just as much as anyone else. But I do believe that there is a benefit to being in the office, more often than not, the earlier you are in your career and perhaps *more so* during your first year in data science.

Recall that one of my tenets is that data science is a learned profession. You simply must rely on the people already working in the field to help you along your learning curve. That process is best learned - and perhaps learned faster - when you are working with someone in the office, shoulder to shoulder. No matter how smart they are, no matter what school they went to, a surgeon is only going to know so many procedures coming out of school, the rest they learned in the field. Would you want a surgeon to operate on you if they learned these procedures remotely?

Another way you might benefit from being in the office is the learning curve that you are going to have to go through to learn the team's current projects. It is a rare occurrence when you can join a team and have a brand new project just waiting for you to get started.

It is much more likely that you will have to be part of a project that is already in flight. You will have to understand the use cases, read the documentation, and study the code. Your team will be looking for you to make adjustments to their existing processes, not the other way around.

Later in your career, you should feel more comfortable asking for the flexibility to work from home. This is all after you've grown as a professional, have a few delivered projects in your work history, and have an established workflow. But that's all a few years from now and maybe you'll be at a stage where you want to be in the office more because you will be mentoring new data scientists too!

Having said all that, I realize that you might live in an area, or even an entire country, where an in-office role for data science is not possible. While I would not consider this an ideal situation, with some extra work and discipline on your part, you can still find success as a remote data scientist if that is your only available option. You will have to become a bigger part of online communities and have more discussions with people around the world about the work you're doing, but it can be done.

Pair Programming

If you're not already familiar, pair programming is the practice of putting two code engineers together to work side-by-side on the same problem or feature. It comes from the world of software engineering where it's popular among some teams. More specifically, it's a concept that was introduced in Xtreme Programming, a form of Agile development.

There are many benefits to pair programming and I'll leave it to you to look up the Xtreme Programming books for that full review. You don't have to use all of this system, or even Agile, to benefit from pair programming. Here are the biggest benefits that I have seen:

1. **Common Code Ownership:** If you have multiple people working on the same code then there is more knowledge spread among the team. There should rarely be a scenario where code can't be changed or updated because someone on the team is on vacation or otherwise unavailable. All code would have at least 1 other person who

knows what's going on and should be able to assist.

2. **Avoids Spaghetti Code:** Working with another person next to you forces you to think through your code and talk out loud. It's amazing how this simple process can help you refine your code and maybe just take that extra few seconds to think about how it can be done better.

   This is all in addition to the feedback you'll also be getting in real-time from your pair programming partner. Said another way, the code you write during pair programming just feels "cleaner" because it's been through both the thought and spoken process by two different people. This carries over into code that has better structure and is usually more readable and understandable than code written by a single person, not talking out loud.

I have used pair programming among my teams and have found great success with it. Here's how I have found it to be most useful:

1. **Interns Helping Interns:** I will usually get at least 2 interns during the summer session. They will come from different schools and backgrounds and have varying skills when it comes to programming. Putting two interns together has been an effective way to have them teach each other as well as produce something meaningful at the end of their time within the organization. While you may not be joining a company as an intern, during the early portion of your career it will help you to be paired up with someone else as you develop your first few pieces of professional code.

2. **Team Building & Collaboration:** If someone on my team is stuck on a significant problem, or maybe they are working on something particularly

notable, I might have them present a piece of code during a team meeting. Then we, as a large pair programming team, will each contribute some code while talking out loud on video. It can be quite spirited to have four or more people working on the same piece of code.

Also, for me as a manager, it can help me gauge the programming skills of the team. I have held these sessions where some of my direct reports have shown me some amazing skills with (as an example) Python contractions or lambda functions that have left me very impressed! This exercise can make for fun meetings.

3. **Sharpen The Saw:** As you grow into a data science leadership role, you will find yourself doing less & less of the daily technical work. So I will sometimes ask to do pair programming with one of my data scientists just so that I can keep my mind fresh with code and concepts. I may be out of practice, but I still know enough to be able to help and guide someone else while also "sharpening the saw" and making sure that I still keep my technical skills.

I completely understand that coding in front of another person can induce a bit of anxiety. I will also admit that I was resistant to pair programming when I was first introduced to it in the early 2000s. But I truly do believe that it has its place in data science and I hope that you will also incorporate it into your practice.

Model Review

I know what you're thinking. "Model review? I've already been through that in school with each professor!"

But, no, this is a whole other kind of review. Whether this is an easy exercise or a difficult one will depend on your organization and the types of controls they have in place.

Essentially, whenever algorithms are deployed, there is an independent team that may come in and ask you to document your model for their review and they will put you through a process to examine the algorithm and give their sign-off. The same process is usually required when you update your algorithm in the future.

In some cases, your company may bring in an outside company - like a consultant or auditor - who will be your review team. No matter who does it, you usually need their sign-off before any deployment to production can take place. Again, there are many aspects to a model review process, each company tailors it for their needs and the requirements they are trying to fill.

It can be daunting to have another person come in and start to take a microscope to your approach and results. As a manager, working through this process with my teams is one of the hardest things I do. The following are some items that I share with my teams to help us all navigate the process.

- **It's driven by policy:** In an ideal world, all algorithms would be perfect, work as desired, and never need any review or maintenance. But that's not the world we live in and, as a result, policy-makers within your organization have put these review requirements in place. In a lot of cases, that policy is also being driven by other forces such as laws and regulations in your particular part of the world.

- **You're on the same team:** The review process can be frustrating and feel like an impediment to progress. But you always have to keep in mind that the review team members are also your colleagues and both teams are ultimately working towards the same goal.

- **It's rationale, not results:** This one is going to vary greatly and you should adjust to whatever you're advised to do. But most of the review teams that I have been a part of are more about presenting your *rationale* for a decision rather than the *results* of the decision. Given what we (should) know about variance, that is a sound decision, and you shouldn't be shy about presenting the results you have, no matter what they are. Just be prepared to back them up and/or contrast your results to whatever your baseline model is.

- **Defend when needed:** It is completely OK to disagree with the review team. If you feel strongly about your rationale and approach to an aspect of your algorithm, then speak up. In any review process, there are going to be some "negotiations" that happen over time. Be prepared with defensible arguments that support your initial approach and defend your stance when needed.

I always tell my teams that the review process requires both honesty & patience; I advise that you do the same. But this process is something that you will have to adjust to during your first year in data science, there is no equivalent that you would have come across during the academic portion of your career.

CHAPTER TEN

Get Out There

OK, time to put this book down!

It is my hope that by reading this book you now understand the practical side of actually *doing* data science instead of just talking about it in a classroom. At some point, you just have to get out there, put the work in, and start to build your professional (not academic) skills and your reputation as a scientist.

If you have some time to prepare prior to joining a team - or maybe you're still looking for a role and interviewing - this is a good time to create your own studies. Get out there, read some articles, use a critical eye when presented with numbers, seek out raw data, put that data through your own models, and form your own conclusions.

You don't have to win a Nobel Prize, no one even has to read any of your work. You just need to start putting yourself in that data scientist mindset and do the practical side of data science. Not only will you be better prepared for your first day, but you will also feel better mentally because you will have already done the same work that you will be asked to do in the workplace.

Don't be afraid, don't be intimidated. Yes, data science can be difficult at times, yes, you will have to work hard, maybe even work late on some days. But data science is also filled with some really good resources that can help you. There's always an opportunity for you to approach a problem from a different angle, create a new view, and examine a new parameter. Most importantly, data science is also filled with good people and good leaders who can help you along and provide some guidance.

If you have enjoyed this book and my approach to data science, I invite you to follow me on all the major social media platforms. I put out content on a daily basis, host live streams and, of course, there are also data science memes! You can find me as @dswithdennis and I look forward to connecting with you in the future.

Finally, I also want to plant a long-term seed in your mind. Earlier I mentioned that I wanted to write this book to help future generations of data scientists as a way to give back to this field that I love. When the time comes, much further along in your own career, I hope that you too will also take the time to reach your hand back to help the new scientists in the field start their careers.

While I don't think data science is in danger of dying out or anything like that, it is important that we teach each other and help the next wave of scientists be better than ourselves. That is how we can all ensure the long-term health and advancement of the profession as a whole.

I'll see you next time.